## ON THE HUNT WITH
# EAGLES

SANDRA MARKLE

Lerner Publications ◆ Minneapolis

# THE ANIMAL WORLD IS FULL OF PREDATORS.

Predators are the hunters that find, catch, and eat other animals—their prey—to survive. Every environment has its chain of hunters. The smaller, slower, less able predators become part of the prey for the bigger, faster, more cunning hunters. But everywhere, just a few kinds of predators are at the top of the food chain. In nearly every habitat of the world, except Antarctica, this group of top predators includes one or more kinds of eagles. Scientists currently list sixty different kinds of eagles. Two of the largest are the Steller's sea eagle and harpy eagle. Steller's sea eagles are found along seacoasts and rivers of Japan, Korea, and eastern Russia. Harpy eagles live in tropical rain forests from Mexico to northern Argentina.

**WOW!**
Large sea eagles have an 8-foot (2.4 m) wingspan! Harpy eagle wingspans can stretch 6.5 feet (2 m).

Why are Steller's sea eagles and harpy eagles top predators? For one thing, the adults are big enough that no other predators regularly hunt them. Females of both kinds of eagles are nearly twice as big as the males and weigh as much as 11 to 20 pounds (5 to 9 kg). These eagles also have giant wings and amazing lifting power. A sea eagle can fly with a fish that weighs up to 8 pounds (3.6 kg). A harpy eagle can carry a 9-pound (4 kg) monkey.

STELLER'S SEA EAGLE

HARPY EAGLE

Even though they're big, sea eagles and harpy eagles are light enough to fly. That's because their skeletons are made of mostly hollow bones. The eagle's coat is made up of more than seven thousand feathers, but the coat is also light. The combined weight of one of these eagle's feather coats is only about 21 ounces (586 g)—less than the weight of two average cans of soda.

**WOW!**
Eagle feathers look solid, but each is really made up of many strands. Each strand is edged with tiny hooks that zip them together.

The amazing flight of Steller's sea eagles and harpy eagles is another reason they are top predators. A Steller's sea eagle's giant wings help it glide through the air with very few wing pumps. This helps it save energy while it searches for prey. Once the eagle picks a target fish, it pulls its wings in close and dives. Nearing the water, it spreads its wings to brake and catch its dinner.

**WOW!**
Bumpy bottoms on a Steller's sea eagle's feet give it a slip-proof grip on fish.

STELLER'S SEA EAGLE

7

A harpy eagle perches on a tree branch until it spots target prey. Then it flies. Its wingspan is just narrow enough for swooping around trees and between branches in the rain forest. Its long tail helps it steer. Once close, it dives to catch its meal.

Another reason these eagles are top predators is their clawlike nails, or talons. Talons are great weapons! An eagle catches prey by stretching out its legs and grabbing with its feet. Its toes—three facing forward and one backward—clamp down, stabbing the prey with the knife-sharp talons.

HARPY EAGLE

9

The harpy eagle's talons are long, but the curved back talon on each foot is giant—as much as 5 inches (13 cm) long. That's longer than the average grizzly bear's claws. A harpy eagle also has such a strong grip that it can crush bones.

HARPY EAGLE

STELLER'S SEA EAGLE

All eagles have strong beaks. These are hard bone covered with a thin layer of keratin, a material similar to human fingernails. Eagles need strong, hard beaks to kill prey and pull off chunks of meat to gulp down. The Steller's sea eagle's massive curved beak is even strong enough that during the winter, it can feed on the frozen remains of dead animals.

HARPY EAGLE

**WOW!**
A harpy eagle can go as long as a week between meals.

A harpy eagle usually catches its prey in the rain forest canopy. It swallows small prey whole. When it eats bigger prey, such as sloths and monkeys, it tears off and swallows chunks. Then its digestive system goes to work.

Food passes into its stomach where strong digestive juices start to break it down. From the stomach, the food passes into the gizzard, a muscular sac. There, the food is turned into a soft mass and passed into the intestine to complete digestion. Any hard waste bits, such as some bones or hair, stay in the gizzard. Hours later, the eagle brings up a pellet of those bits, opens its beak wide, and lets this waste drop out.

Another reason eagles are top predators is their keen eyesight. Eagles have large eyes with super magnifying lenses for excellent long-distance vision. They also hunt during the day, when it's easier to see. When light enters an eagle's eyes, it strikes the retina, a layer at the back of the eye packed with light-sensitive cells. Any of the cells triggered by light send signals to the eagle's brain. Once the brain interprets these signals, the eagle sees.

STELLER'S SEA EAGLE

NICTITATING MEMBRANE

STELLER'S SEA EAGLE

The feathers around an eagle's eyes act as eyelashes, providing shade and keeping out dirt. In addition to the upper and lower eyelids, eagle eyes have a third eyelid, the nictitating membrane. This slides across each eye. It cleans the eye and, like goggles, it covers and protects the eagle's eyes from windblown dirt.

A harpy eagle has a special adaptation for finding prey—a disk of facial feathers. When searching for prey, that disk picks up sounds and funnels them to its ear openings. By evaluating whether its left or right ear hears the sound louder, the eagle homes in on prey it doesn't see.

Of course, both harpy eagles and Steller's sea eagles hunt to eat and stay healthy. But a time comes in adult eagles' lives when they hunt for another reason—to raise eaglets.

HARPY EAGLE

In April or May, harpy eagles that are about five years old find a mate for the first time. They'll remain together as long as both live, but they only nest to produce young every two to three years. That's because it takes them that long to raise their eaglet. They provide food for it until it's able to hunt for itself and leaves their territory.

A pair that's ready to nest chooses a site high up in a tree's branches. That location keeps the eaglet safe from other rain forest predators, such as jaguars. They may only repair an old nest. But if building a new nest, the pair carries branches to the site with their feet and arranges them.

Once the nest is complete and the pair mates, the female usually deposits two eggs. She sits on the eggs while the male hunts and delivers food to her. The eggs need to be incubated, or kept warm so the eaglets inside will develop. The female stops incubating when one eaglet hatches, so there is only one young to be cared for. Raising this youngster will take both parents a lot of time and effort.

**WOW!**
Both harpy eagle parents walk across their nest with their feet balled into tight fists to protect their eaglet from their sharp talons.

HARPY EAGLE

## WOW!

A Steller's sea eagle nest is about 6.5 feet (2 m) wide and just over 3 feet (1 m) deep. Scientists studying sea eagle nests found they may be made of more than four hundred branches.

STELLER'S SEA EAGLE

In February or March, any Steller's sea eagle about six years old or older chooses a mate. The couple may have paired up before, but sea eagles often choose a new mate each breeding season. Then they may repair and reuse an old nest. If they are building a new nest, they choose a big tree or cliff site close to where they can easily catch fish to feed themselves and their young.

After mating in April or May, the female produces one to three eggs. The female spends the most time incubating the eggs, though the male and female share the job. The eaglets hatch after about forty-five days.

The newly hatched harpy eaglet is teacup-sized with a thin coat of down, or fluffy feathers. Its mother stays close to keep it warm and shelter it with her wings when it rains. Meanwhile, its father continues to deliver prey to the nest. For about the first three weeks while the eaglet grows stronger, its mother tears off small pieces and feeds it.

By the time it's forty-five days old, the harpy eaglet is much bigger and stronger, and it's developing adult feathers. At around five months old, both parents are hunting to bring it food. And the nearly adult-sized juvenile is branch walking and flapping its wings to get ready to fly. But when the juvenile first launches from a branch, its flight is shaky, clumsy, and very short. It needs lots of practice to fly well and successfully hunt for itself.

HARPY EAGLE

STELLER'S SEA EAGLE

Steller's sea eagles often hatch and raise three eaglets. At first, the youngsters only weigh about 3 ounces (85 gm)—about as much as thirty US pennies—and only have a thin down coat. The eaglets' parents take turns staying with them to keep them warm and feed them bits they tear off fish. But soon the eaglets are big enough to stay warm on their own and beg loudly and often to be fed. Although there's danger of crows or martens attacking the eaglets, both parents hunt to feed them.

Ten-week-old Steller's sea eaglets are nearly adult size and have adult feathers. They prepare for flight by hopping and flapping. When they're strong enough, they make their first flight.

STELLER'S SEA EAGLE

## WOW!
Juvenile sea eagles usually have a brown coat, which is different from the adult's coat color. But juveniles have the same bright yellow beak.

A juvenile harpy eagle remains in its parents' territory for two to three years. When its early hunting attempts fail, it returns to its nest and its parents deliver food. Such extended care gives the juvenile the best possible chance of becoming a top predator in its rain forest habitat. Then one day, it will produce yet another generation of harpy eagles growing up to be top predators.

HARPY EAGLE

Shortly after its first flight, in August or September, a juvenile Steller's sea eagle leaves its family's nest. Sea eagles often gather in groups while hunting. That lets young sea eagles mimic how adults hunt.

Once a Steller's sea eagle has been a successful predator for about five to six years, it will claim a mate at the next breeding season. Then it will share in hatching and raising another generation of top predators for its habitat.

STELLER'S SEA EAGLE

# A NOTE FROM SANDRA MARKLE

**Even top predators face challenges**, and a global change in Earth's climate is challenging both Steller's sea eagles and harpy eagles, though in very different ways. Warming Arctic conditions keep ice melted, providing access to fish for sea eagles for much longer periods each year. While that could be viewed as a positive for sea eagles, the lack of sea ice, especially thick ice remaining year-round, is already leading to increased commercial fishing, decreasing the sea eagle's food supply. There has also been an increase in oil drilling that could cause pollution and kill fish. Plus, oil on the water can coat sea eagles, harming them.

Global climate change means less rainfall for the harpy eagle's rain forest home. Over time that will cause a decline in plant growth, making it easier for these eagles to find and hunt their prey. However, fewer and less healthy plants will cause a decrease in the number of the plant eaters harpy eagles hunt. Eventually, the harpy eagle population may also shrink.

Photo by: Skip Jeffery Photography

# SNAP FACTS

## STELLER'S SEA EAGLE

### ADULT SIZE
Females have bodies up to 3.4 feet (1 m) long—from head to tail tip—with about an 8-foot (2.4 m) wingspan, and weigh as much as 11 to 20 pounds (5 to 9 kg). Males are smaller.

### DIET
They mainly eat fish but also often eat crabs, shellfish, and small animals.

### LIFE SPAN
In the wild, they may live as long as twenty-five to thirty years.

### YOUNG
The female usually lays one to three eggs sometime from April into June. For about forty-five days, chicks develop inside the eggs before hatching. Then chicks remain in the nest for about another thirteen weeks while developing to be ready for flight.

### RANGE
They live along seacoasts and rivers of Russia, Japan, and Korea.

### FUN FACT
Scientists studying Steller's sea eagles have recorded them diving at 30 miles (48 km) per hour.

## HARPY EAGLE

### ADULT SIZE
Females are larger than males and may be 3.4 feet (1 m) tall, have a 6.5-foot (2 m) wingspan, and weigh as much as 11 to 20 pounds (5 to 9 kg). Males are smaller.

### DIET
They mainly eat tree-dwelling animals, such as monkeys and sloths.

### LIFE SPAN
In the wild, they are believed to live as long as thirty-five years.

### YOUNG
One to two eggs are laid and incubated for about fifty-six days before hatching. As far as scientists studying these eagles know, only one eaglet hatches, is raised, and continues to be fed near the nest for as long as its entire first year.

### RANGE
They live in tropical rain forests from Mexico to northern Argentina.

### FUN FACT
Harpy eagles can spot something less than 1 inch (2 cm) in size from as far away as twice the length of a US football field.

# GLOSSARY

**BEAK:** an eagle's hard, sharp-tipped jaws

**DOWN:** fluffy feathers that are an eaglet's first coat

**EAGLET:** a baby eagle

**EGG:** the hard-shelled structure within which a baby eagle develops

**JUVENILE:** a young eagle that is adult size and has its first coat of feathers

**NEST:** a structure an eagle builds or chooses to lay and hatch its eggs and raise its young

**PELLET:** a compacted ball of waste from digestion, including bones, that the eagle brings up

**PREDATOR:** an animal that hunts other animals

**PREY:** an animal that a predator catches to eat

**TALON:** an eagle's claw

# INDEX

beak, 11, 13, 25

digestion, 13

eaglets, 17–18, 21–22, 24–25
eyes, 14, 16

feathers, 5–6, 16–17, 22, 25
flight, 4–6, 8, 22, 25, 27

habitat, 2, 26–27

mate, 18, 21, 27

nest, 18–22, 26–27

predators, 2, 4, 6, 8, 14, 18, 26–28
prey, 2, 6, 8, 11, 13, 17, 22, 28

skeleton, 5

talons, 8, 10, 19

weight, 4–5, 24
wingspan, 4, 8

Image credits: Osamu Asami/Getty Images, p. 3 (left); Barry B Doyle/Getty Images, p. 3 (right); Harry-Eggens/Getty Images, p. 4; Nature Picture Library/Alamy Stock Photo, pp. 5, 9, 11; Michel & Gabrielle Therin-Weise/Alamy Stock Photo, p. 10; Agencja Fotograficzna Caro/Alamy Stock Photo, p. 12; Manuel Romaris/Getty Images, p. 15; JonNPoulsen/Getty Images, p. 16; Alfredo Maiquez/Shutterstock.com, p. 17; blickwinkel/Alamy Stock Photo, pp. 19, 23; AlexeyGnezdilov/Getty Images, pp. 20, 24; USO/Getty Images, p. 25; Octavio Campos Salles/Alamy Stock Photo, p. 26.

Cover: USO/Getty Images.

THE AUTHOR WOULD LIKE TO THANK MARTA CURTI, PEREGRINE FUND, BOISE, IDAHO, AND DR. MICHAEL MCGRADY, INTERNATIONAL AVIAN RESEARCH, KREMS, AUSTRIA, FOR SHARING THEIR ENTHUSIASM AND EXPERTISE. A SPECIAL THANK-YOU TO SKIP JEFFERY FOR HIS LOVING SUPPORT DURING THE CREATIVE PROCESS.

FOR SAMMY REBANDT AND ALL THE CHILDREN AT EAST JACKSON ELEMENTARY SCHOOL IN JACKSON, MICHIGAN

Copyright © 2023 by Sandra Markle

All rights reserved. International copyright secured. No part of this book may be reproduced, stored in a retrieval system, or transmitted in any form or by any means—electronic, mechanical, photocopying, recording, or otherwise—without the prior written permission of Lerner Publishing Group, Inc., except for the inclusion of brief quotations in an acknowledged review.

Lerner Publications Company
An imprint of Lerner Publishing Group, Inc.
241 First Avenue North
Minneapolis, MN 55401 USA

For reading levels and more information, look up this title at www.lernerbooks.com.

Main body text set in Aptifer Slab LT Pro medium.
Typeface provided by Linotype AG.

**Editor:** Brianna Kaiser **Designer:** Lindsey Owens

**Library of Congress Cataloging-in-Publication Data**

Names: Markle, Sandra, author.
Title: On the hunt with eagles / Sandra Markle.
Description: Minneapolis, MN : Lerner Publications, [2023] | Series: Ultimate predators | Includes index. | Audience: Ages 8–12 | Audience: Grades 4–6 | Summary: "Eagles' feather-covered wings, sharp talons, and keen vision make them the top predators in their habitats. These excellent hunters can easily spot prey and glide through the air. Learn about the lives of eagles"— Provided by publisher.
Identifiers: LCCN 2021054008 (print) | LCCN 2021054009 (ebook) | ISBN 9781728456300 (library binding) | ISBN 9781728464381 (paperback) | ISBN 9781728462417 (ebook)
Subjects: LCSH: Eagles—Juvenile literature. | Predatory animals—Juvenile literature.
Classification: LCC QL696.F32 M2579 2023 (print) | LCC QL696.F32 (ebook) | DDC 598.9/42—dc23/eng/20211104

LC record available at https://lccn.loc.gov/2021054008
LC ebook record available at https://lccn.loc.gov/2021054009

Manufactured in the United States of America
1-50700-50119-2/23/2022